Frédéric & Pierre Lepeltier

THEORY
I – THE MENTAL REPRESENTATION OF THE UNIVERSE

June 2014

Preface

Time has come to "rethink" the world.

All the pieces of this "new representation of the Universe", which – in the literal sense – also is its metamorphosis, are gathered under our eyes, like a puzzle: All that remains to be done is to put them in order, to assemble them into a "Unified Theory".

The basic principles of this Theory, a heralded goal, approached but never reached through a scientific process, are laid down today.

- *Establish a coherent, complete, and dynamic model of the Universe*

- *Produce the unified "Code" of Matter, Living, and Psyche*

- *Perform the expected bond of scientific and mythical thought into a new Reason*

And thus, establish (for a brief moment) the time of the "finite world".

We call "crisis" the current situation of the world because it escapes (that's the least one can say) our will. We are subjected to it and our accumulated technological victories do not let us forget the unconquered riddles. Taming reality, making the future "legible", would bestow us with increased power, the power to decode the future and to invest our efforts in prescience.

Reason such as we remember it from the heritage of the Elders and first and foremost from Aristotle, such as it has been inflected, enhanced – and at the same time diminished – under the influence of scientific thought, never ceased to exclude from its field entire sections of the psychic production.

It's as if, to better understand and conquer the world, it had to abandon its native ground and burn its vessels: The field of "irrationality" is more extensive today than it was yesterday. Some even go as far as believing that "the world has gone mad".

5

This is the sole cause of the vague, but omnipresent, feeling that mankind is lost, incapable of projecting itself into the future.

The time has thus come to re-annex these territories to our modern conquests, to reconstitute the Empire of Reason in its true dimensions.

The structure of human rational knowledge has, over the past few centuries, moved away from its roots to the point of renouncing them and, thereby, undermined both itself and the fantastic adventure of humanity.

The bases of the - paradoxically called - "fundamental" sciences are certainly not obsolete, but the scientific structure has more "fundamentally" lost its foundations.

This has concededly not stopped us from conquering the Moon and aiming for new planets, or from designing ever more sophisticated machines, systems, and networks, nor has it troubled our daily lives.

However, as its structure now stands, our knowledge will not allow us to uncover the secrets of matter, or grasp the mysteries of life and, even less so, to penetrate the depths of the human's brain or psychological activity.

No matter how hard and how good scholars of all fields add new layers to humans' knowledge, search ever further into the infinitely large or infinitely small, they just further weaken this paltry tower, and exhaust themselves, not in the pursuit of the Holy Grail but of a growing uncertainty, fueled by an inextinguishable thirst that only grows at every sip.

Gullibles, conformists, and those – including today's decision-makers – who have left science in the hands of experts are blissfully impressed by its growing complexity. At a time when Moliere and Andersen's "The King is Naked" resonates, the reign of preconceived ideas is unfortunately alive and well.

As science increasingly turns its back on knowledge, it is hardly surprising that all levels of social, economic, and intellectual activity of human society reflect a worrisome uncertainty.

The authors' paradigm is not new. Plato developed 2,300 years ago, in Timaeus. Closer to us, sixty years ago, Erwin Schrödinger said :

"Matter is an image in our mind – the mind therefore comes before matter" (Science and Humanism: Physics in Our Time).

Great geniuses, such as Einstein, Heisenberg, Schrödinger (and a few others with the same intuition) have this unique ability to show the way to the next generations, with generosity and discernment.

We have tried to answer their call with the respectful ingratitude of children who claim both legacy and originality, seeking to accomplish our share of this thrilling work of decoding the enigma.

Building on all knowledge accumulated over the past decades and weaving together mythical or 'savage thought' with scientific reflection, we propose to expound the paradigm as follows :

Our representation of the world, inherited from our roots in matter and from all our successive metamorphoses, in turn structures the same world.

No universe can be conceived, perceived or even imagined unless it has first taken shape in the mind. The subject, the observer, this outcast of the fundamentalist scientific thought who interferes with the experiments and challenges the stages of matter, is, in the end, the sole and true seat of knowledge.

The mental perception of the Universe contains all that which one can perceive, conceive, imagine, of this Universe in the eternity of its duration and to its infinite extent.

It is for this "perfect model", the only one that can make us an "omniscient observer" (as described by Laplace), that we must strive.

> "We can consider the current state of the universe as the effect of its past and the cause of its future. An intelligence that should know at a given moment all the forces that put nature into motion and all the positions of all objects that nature is made of, if this intelligence in addition were sufficiently ample to submit this data for analysis, it would hold in a single formula the motion of the largest bodies in the universe and of the smallest atoms; for such an intelligence nothing would be uncertain and its own future as well as its past would be obvious to its eyes" A Philosophical Essay on Probabilities,
>
> Pierre-Simon Laplace, ca. 1820

Therefore, it is this "mental representation" that we propose to examine.

Early in this audacious search, one will perceive not the Universe, but its primitive and mute image, the shapes and numbers, proportions and har-

7

monics, chaos and order, motion and forces, energy and matter, space and time, gravity and radiation and, ultimately, Life and Thought and all that is the Matter, the Living, or the Mind

All components of the Universe – or at least their "possibility" – are already there, in the human's mental representation.

We also intend to demonstrate that this "representation" harbors the entire set of Laws, principles, and interactions that govern each of the material, living, and psychic components of reality.

These laws and principles, sometimes built up into seemingly disparate "scientific theories", all result from the "structure" of the primitive representation of the Whole.

This is the matrix of all knowledge and the source of all principles and of hypotheses that the human's understanding can apprehend, expound, comprehend, and demonstrate.

It is the only place where these perfectible theories can find their unity and truth. It is the arborescent structure of knowledge or, the sole and true "Tree of Knowledge".

INTRODUCTION

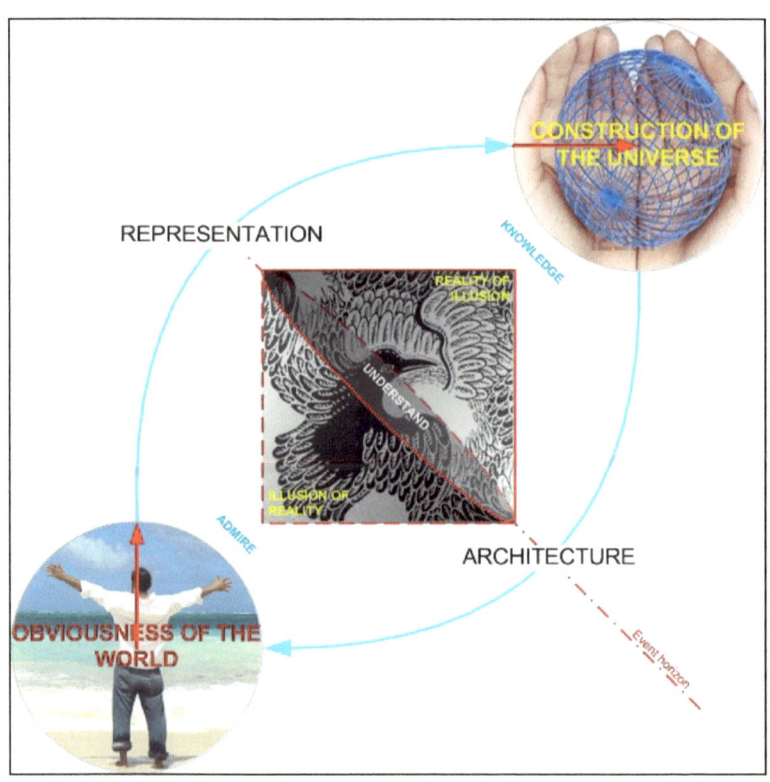

Everything should be made as simple as possible but not simpler.

Albert Einstein

The Copenhagen controversy, the ultimate artificial confrontation between idealism and realism opposing the fathers of atomic physics to those of quantum physics, has an epilogue with the reconciliation brought about by the recent effective detection of the *massive scalar boson*, the last element of matter whose existence was stipulated by Peter Higgs 50 years ago.

Once again the experiment "certified" the result which was in fact *already* contained within the mental construction of the Universe: "matter is in the image of our mind – the mind is therefore anterior to matter" (Erwin Schrödinger, *Science and Humanism*.

A new era of Knowledge opens up before us which could almost be described as "post-scientific". We may legitimately call this structured mental representation – in the sense of a "world system" of Galileo, Newton, Laplace, or Duhem – a *model*, and it may thus be used as the *unique, complete, and certain* source of knowledge.

This model, freed from the constraints of causality, contains latent within it all the components of the Universe (however imperceptible they may be), its dimensions and forces, as well as the architecture of its Laws.

It reveals the full set of possibilities as well as the fractal matrix not just of states and events which have already come about but also those of the future – it is an "open cast mine" for future discoveries.

This revolution in our way of exploring and understanding the Universe and our world deriving from it, together with other major transformations in the "human phenomenon", is opening up the new space of a "knowledge economy" – far beyond our historical perspective and the *crisis* obscuring its immediate horizon.

But there is a final detail that needs sorting out – or rather a preliminary. Despite the progress made by the neurosciences, cognitive sciences, cybernetics, and information technology – and even advances in the various disciplines examining the psyche (and doubtless because of the compartmen-

11

talisation of all these various disciplines) – nothing has been said about the process by which this representation of the mental "model" of the Universe is actually formed.

And it is precisely this latest experimental exploit, this discovery of nothingness, this culmination of science that brings us back to the inspiration that there is indeed a unified theory of the Universe, that this theory is simple, and that the question behind it is: *"how do I see the world ?"*

The Failure of Scientific Reasoning

In favouring distinction as the first course of action and measure as the unique form of reason, the quest for knowledge, become 'science', has dedicated itself to dividing the All indefinitely, in the illusory search for a unity on which it turns its back.

Thus, it has condemned itself to dissecting indefinitely the path of an arrow, thereby condemned to never being able to determine simultaneously its position and its speed, whilst still knowing that it will reach its target.

It has itself defined distinct fields of study, including for each discipline specific dimensions and forces, but surrounded by a void which is impossible to bridge to reunite the pieces.

Each discipline is consequently like a god of the Pantheon, Poseidon or Neptune for the ocean, Mars or Vulcan for fire, too proud of their own attributes to cooperate in the building of a unified world.

By distinguishing each part of the world from an untitled void, by expelling the onlooker from the Universe, despite it containing him, and placing observation in a position of absolute primacy, science forces itself to integrate definitively the illusory dimensions of time and gravity, and in corollary, entropy and energy.

As a consequence, despite continuous technological progress that applies ever improving means of observation, the scientific process is nothing more than an iterative sequence of apperceptions of the world tested in laboratory experiments.

Incapable of accounting for the function that presides at this fractal sequence, lost in complexity despite the probabilistic and statistical artifices it uses to justify the discrepancy between its measurements and the successive states of the world, science does not succeed in providing us with a homogenous representation able to *establish a unified model of the Universe*.

13

The Limits of a Realistic Representation

Modern science is the heir to empiricism and is unswervingly attached to the certification of its hypotheses by experimentation, adhering to what it calls reality whilst at the same time undermining it.

From the moment of the very first observation in a scientific investigation, the *reality* of the object phenomenon is established as the yardstick against which all the stages in the reasoning and ultimately the scientific law itself is to be compared.

Even though the constant evolution of the manifest world (a corollary of time and gravity) make it impossible for phenomena to identically recur naturally, science still seeks to reproduce them via experimentation.

Science is thus continually sent back to its errors, and tirelessly repeats the same process right down to the slightest degree of uncertainty so as to take these errors into account.

A scientific law, established on this illusory basis, merely isolates the common elements in the generation of all these phenomena, automatically excluding occurrences of exceptional dimensions, and considering them as statistical artefacts even.

Whereas rationally exceptions should disprove the rule, they are instead pompously declared to confirm it.

By taking observation as the prime form of investigation and furthermore according it primacy, and by taking the confrontation with the uncertain reality of the world as the condition for scientific reasoning, science and its laws fail to account for the full set of possibilities in a "realistic" way.

Being always one step behind the world science can only give an "idea" what the world is without ever being able to reconstitute it. It will therefore never arrive at a coherent structured representation which could underpin *a world system.*

Incompatibility between Models

Given these preconditions it is hardly surprising that science finds itself up against a "quantum wall", beyond which measurements merely reflect the sole certainty that there is an observer present. The theories for both spatial and temporal quanta are limited, and in seeking to attain an impossible state of perfection the theories tend towards ever greater and ultimately paralysing complexity.

Torn between the need to distinguish between first and final units, waves and particles, speed and position, scientific reasoning finds itself constantly obliged to build its models on the basis of two incompatible principles.

The first of these principles, in a vision derived from the world, privileges speed over position and makes it possible to establish (though shrouded in uncertainty) all of the *possible events* in the world (beyond the limits of observation). But the counterpart of this is that it is unable to discriminate between one possibility and another one.

The second of these principles on the other hand offers a complete vision of all of the manifested possibilities (*states*) and of each of the movements from one state to the next. But whilst it determines the set of what is observable, it does allow us to go beyond the limits of the observable or to discover universality.

Irrespective of the scale of means used to push back the limits in observing the world, science finds itself caught between a universal model unable to determine anything at all, and a determinate model unable to account for the Universe.

Pretending to ignore that distinguishing the world from the Universe presupposes nothingness, and that observation presupposes an observer, scientific reasoning finds itself wholly unable to account for the unifying principle. However revolutionary or appealing its models may be, they are necessarily *incomplete models*.

15

The New Paradigm

The inability of science to unify the Universe and the world (its manifest derivative) means we need to redefine our sources of knowledge about them – to produce a new paradigm.

The precondition for such a paradigm, and which does not call into question the primacy of observation, holds that the observer together with the world in which he is present constitute the observable part of the Universe, that nothingness is the non-observable part of it, and that both are necessary for it to exist.

Nothingness, which according to scientific reasoning renders some of its theories unverifiable, thereby becomes the set of possibilities not yet made manifest and the alternatives to possibilities made manifest using the same elements but in different combinations.

The new paradigm therefore needs to take account that our observable world is not order distinguished from disorder, nor something that has emerged from nothingness, but one order amongst others and one world amongst possible worlds, all of which are derived from the same universal origin.

A unified model built on these bases would reflect the full set of elements reduced to their simplest existence, and thus to the state of greatest adaptability, and structured in such a way as to account for the set of both observable and non-observable possibilities.

Such a model would always make it possible to distinguish between observed possibilities within the world, as well as making it possible to distinguish them from those which it is impossible to observe. Either because they are part of the unobservable, or because they are part of what has not been observed but which we realise necessarily came about for a given event to have been observable.

It would account both for all the speeds and all the positions of the path of the arrow from the bow to the target. It would be *perfect*.

1. EVIDENCE OF THE WORLD

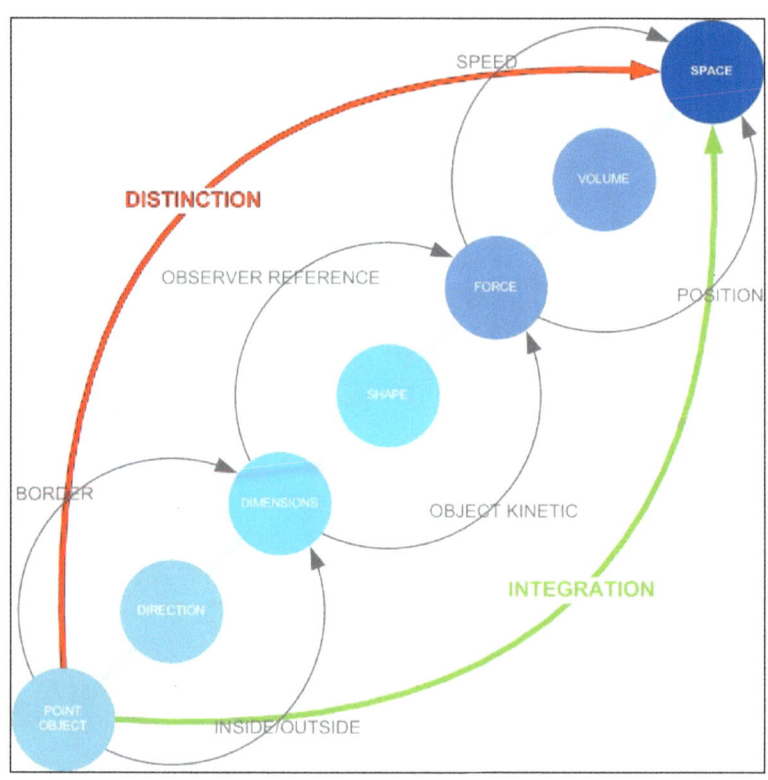

The "world" is evident to us, but it would be foolhardy to endeavour to say what that world is without first defining the conditions of its evidence, something which we shall do here.

We can at least try to say what we know.

The world is made up of the set of manifest objects which we can directly perceive. This supposes that these objects are distinguished from one another and from the world. The world is the primal space where objects are manifest. It is the manifestable All.

Our first objective, as in painting or sculpture, is to define the common supporting *medium* from which the world or planet, the molecule or atom, or indeed any known set together with all that it may contain are distinguished.

This space is also that of the manifestation and representation of these objects. It is the space of universal manifestation.

This supporting medium without which the world would be unknown to us is made up of "elements", where this word is to be understood in a way analogous to that used by physicists to designate elementary particles – but used here without presupposing the existence of any particle.

These elements are present in the space of manifestation and space of representation, but nothing at this stage guarantees that they are the same as those which presided at the creation of the world.

They are the elements of our experience of the world, not the building blocks constituting that world.

They therefore comprise the set of ways in which the world is perceptible to us.

As this space is the supporting medium common to all objects and individuals who perceive and represent it, we can be sure - and it is already a significant result – that these elements enable us to universally perceive the world and conceive it in the same terms.

They are therefore the elements of the space of universal perception.

Were this not the case we would be entitled to speak of multiple worlds where each person would apprehend the world according to their own means, and there would be as many known worlds as individuals to which it was manifest.

Hence our objective is to establish the bases for the *universal protocol for experiencing this world.*

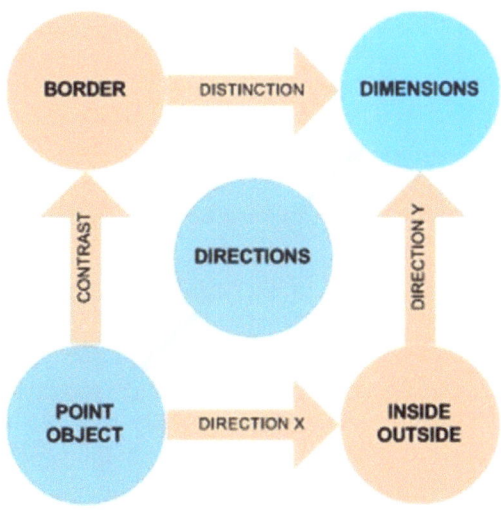

Dimensions and directions

An object appears to us in an evident manner because it is distinguished from a medium, even if this medium is invisible. But all we can say at this stage is that there is something and nothingness. Basically that there is a point or dot on a blank page.

Unless we accept that this dot or point represents the world and everything it contains, we need to further develop our acquisition of an image of the world.

It is in seeking for a universal means to account for the characteristics which distinguish this point from any object that we make our first "discovery" about the elements of the world – dimension.

An object, once it has been distinguished, contrasts with nothingness in the way a dot does on a blank sheet of paper. There is an invisible, fixed, and determinate border constituting an envelope separating the object from its medium – though we will not yet seek to describe its actual form. The two sets and this border make a whole.

This line, which we apprehend at the same time as the object, is one state variable amongst the many characterising it, it is the first *dimension*.

21

"First" is not to be understood here as designating the commonly employed order of length, width, and height - it is rather the first to be perceived, especially as we cannot yet state how many dimensions there are, even less name or order them.

Establishing the first characteristics of the object amounts to following this frontier so as to describe it. That presupposes subdividing it into sections, with the number of sections required increasing in tandem with the level of precision required.

That presupposes the existence of a unique, invisible reference dimension, which serves purely to generate the subdivisions within all the perceived dimensions. It is universal.

At each of these distinguished points, so-called "secondary" dimensions intersect with the first dimension. There are only two possible types of object envelope covered by this thereby closed network of dimensions.

The first possible type of object envelope is one made up solely of perfectly regular dimensions, giving what could be called a potato-shaped lump. Describing it may involve using any one dimension its envelope comprises, or else may require a description of the infinity of dimensions making it up. The space required for reconstituting the characteristics of this object is either one-dimensional or else infinitely multi-dimensional.

For the second possible type, the object envelope comprises at least one irregular dimension (not necessarily the first), or in other words with a perceptible peak. The edge revealed by the discovery of this vertex makes it possible to group dimensions together on the basis of which side of the edge they are on, where the edge too is a dimension.

This means that the envelope may be subdivided into as many faces as there are sets of dimensions, where these sets are defined by the edges present on the surface of the object. This subdivision is totally independent of that generated by the implicit dimension.

The edges, which are characteristics of envelope, constitute the "frontier" dimensions defining the space required to account for this object, but this object only. These dimensions are *native*.

And so within our experience of the world there are objects characterised by a reasonable number of dimensions, where reasonable here means that there is neither one single dimension and the implicit dimension, nor an infinite number of dimensions including the implicit dimension.

22

An object, like a diamond, whose envelope is infinitely subdivided, can at best limit only to half of infinity the number of dimensions defining the common space in which it would be possible to describe all the objects in the world.

Unless we accept that the objects and the world have as their common medium of representation a multidimensional space, such as a Calabi-Yau space (comparable to a crumpled piece of paper) or else a point, then space needs to be defined using the smallest sufficient number of dimensions to account universally for all objects.

Whenever we consider several objects or else several dimensions within a same object this implies that these objects or dimensions are comparable, that is to say of such a kind as they may be compared and are not identical. In order to "reason" the world we first need to discover the links making them comparable.

These "physical", native dimensions which belong only to objects – unlike the implicit dimension (which we cannot as yet state to be time) – define a space within which objects may be compared.

Since the dimensions of an object differ from one another at least in terms of their direction, this space cannot in any event be one-dimensional. Nor can it be infinitely multi-dimensional, otherwise it would become impossible to establish the common points between dimensions.

This common space for describing all the objects in the world cannot reasonably comprise more dimensions than those representing the most singular directions. This amounts to defining the object with the smallest number of faces possible without the secondary dimensions defining these faces being either too numerous or too similar.

A space thus circumscribed, that as is already clear cannot be based on a complex axiomatic, would thanks to its economy be able to act as the medium supporting a simple, universal system susceptible to account for all objects and for the All.

All evidence bears within itself the instruments for its description.

1a.1 : An object appears because it contrast with the world around.

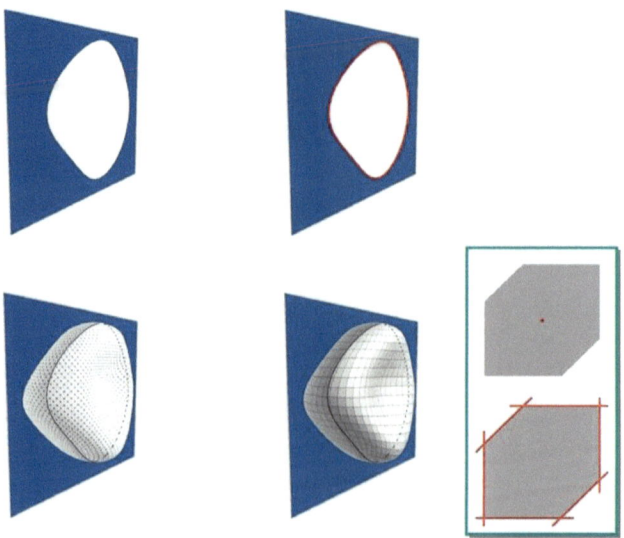

1a.2 : An object's dimensions are discovered after its boundary with the world had revealed itself.

24

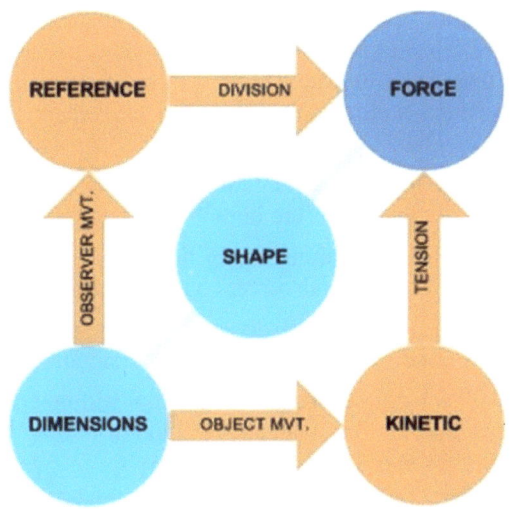

Forces and forms

The set circumscribed by an object envelope is made up of the same elements as the void. The only separation between them is the first dimension attendant upon its evidence. And this object is only differentiated from another object by the different form of this dimension.

This leads us to the discovery of a new element in the medium supporting the world – *forces*.

We cannot distinguish any division within nothingness, unlike the object envelope. This absence of any reference division enabling us to characterise nothingness leads us to suppose that any divisions which may indeed exist within it are unaffected by movement.

Yet when we distinguish an object from nothingness, the dimensions making up its envelope lead us to suppose that their only purpose is to limit any potential movement of its constitutive divisions to the volume of the object itself.

Hence the divisions within nothingness, which are unconstrained by an envelope, may move freely (though not within the place occupied by the volume of the object).

Nothing distinguishes the divisions within the dimensions of an object from the divisions constituting the volume of an object or nothingness other than the fact that they are generated by the implicit dimension. It is only their generation which limits their potential for movement to the direction of the dimension to which they belong.

Unless there is an apparent movement on the surface of an object, we have to agree that all these dimensions are immobile in relation to each other.

There is therefore a "tension" transmitting the kinetics of a division to all its neighbours, equally and in both directions along each dimension.

Force applies universally to all the divisions within all dimensions of all objects, and is thus not constrained to any single orientation.

It is the entire object considered as a division within nothingness that moves or changes attitude. Its potential for motion is only limited by the presence of other objects or observers.

The fact that there is no fixed reference within nothingness means it is impossible to establish whether an object is at rest or in motion. All that we may say is that it may be in one state or the other. Nor is it possible to establish unambiguously whether it is the observer or the object which is moving, and in what direction.

The potential for movement of these divisions originates in force, and it maintains unabetted the cohesion of all the dimensions of all the objects distinguished in the world. It precedes objects. This force is primordial and hence unique.

Because all objects are made up of a limited set of divisions, its volume and evidence are derived from this force. As the combination or disintegration of two objects necessarily follows on from an event brought about by this force, it may be said to have the potential to make any form manifest.

When in motion an object disrupts the divisions within nothingness surrounding it, and these in turn transmit this kinetics. But because these divisions do not belong to a dimension they do not remain motionless in relation to each other. This is chaos.

The fact that this force maintains the cohesion and form of objects means objects cannot transgress the limits imposed on all the other divisions within nothingness, including the envelope of other objects. This force is therefore instrumental both in distinguishing an object from another object and in distinguishing it from nothingness.

26

The fact that there is no fixed reference preceding objects - and nothing-ness cannot play such a role - means that any elements, objects, dimensions, and divisions may be supposed to be at rest whilst at the same time possessing the potential to change attitude and move in all directions equally.

The characteristics specific to the movement of each object may look to us as if they result from different forces. Nevertheless the only effect these forces have on objects or divisions is to set them in motion irrespective of their orientation, without ever breaking the cohesion of the dimensions, and so they are merely occurrences of the unique force.

Whatever the number of any such occurrences, they are all unified by their common source.

Along the orientation of any dimension this force set both the dimension and us in opposite directions. The one we take to explore it and the one the dimension takes, for us to be able to explore it while still immobile. As so, this set of directions of the unique force is substituted by kinetics and po-tential – as Newton discovered with his Apple –an action and a reaction.

The kinetics takes the form of a supplementary dimension similar to di-mension generating the divisions within the object envelope, and which we call *time*.

The potential is diametrically opposed to it and, whilst we are not yet in a position to call it *gravity*, it maintains the cohesion of the world-system as an object-system, and consequently the cohesion of the system common to their representation.

All forms make manifest the existence of the unique force.

1b.1 : The form of an object is restored by its dimensions.

1b.2 : The unique force necessary to discover an object is applied to every object.

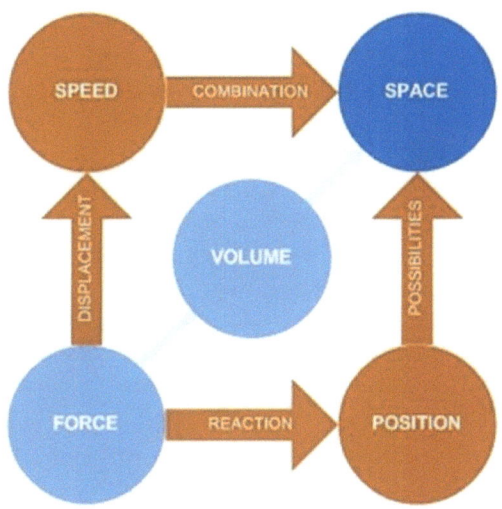

The Space for Possibilities

Adding forces to the set of dimensions is not yet sufficient to constitute a supporting medium such that the world could unfold and we could have a conceptual instance of the world.

Forces and dimensions need first to take on organised form within a construction that renders universal the space in which the world can become manifest.

The "physical" or native dimensions of an object are distinguished as being those which form the set of the envelope of this object. As soon as these native dimensions take on the evidence inherent to the object they are potentially perceptible (and for that matter the evidence inherent to the object coincides with the perception of a first physical dimension, without which the object would not exist).

They are the supporting medium for *assessing* the object.

A force, however, can never be directly perceptible to us. It becomes measurable by setting the physical dimensions of an object in motion. It is thereby transmuted so as to conform to the constraints of our immediate perception. This is the case, for instance, when light is transmuted from imperceptible solar wind to aurora borealis.

The space enabling the world to unfold comprises, in addition to the dimensions required for describing the physical characteristics of objects, two further dimensions which may be substituted for the two opposing directions of the unique force referred to above, and which we normally identify as action-reaction.

Given that any description and representation of the world cannot contain as many dimensions as the world does, only a certain number of dimensions may be used to define the space of this representation.

Whilst the "physical" or native dimensions directly perceptible to us cannot thereby be wholly absent from this space, the fact that we need to substitute a "transmuted" dimension for them in order to describe and represent the occurrences of the unique force indicates the number of dimensions this space is composed of.

The absence of any native physical dimension within the space of representation would render the object impossible to describe and thus impossible to represent. Equally the absence of any native physical dimension within the space in which the world unfolds would have as its corollary the absence of the object or at least the impossibility of its being perceptible.

If however the space of representation included the same number of dimensions as the space required for the object, then the object would become immediately and totally accessible to any consciousness without any need to describe it. This is the case for what is called a primitive geometric form such as a cube. Consequently it would not need to be rendered manifest. The two spaces would be one.

The world imposes its evidence on us, but not its unity. It is from this evidence that we derive the instruments constituting the space within which the world unfolds.

But without a space of representation no consciousness would be able to see any object, even less distinguish it, or conceive of the Universe on the basis of that object.

Yet whenever several consciousnesses perceive the world in the same way, the space within which it unfolds and the space of its representation (which are distinct from one another) become universal.

A point on a line or a dimension is sufficient to represent the presence of an object. To describe the movement imparted to this object by a force in the same space of representation it needs to include a second, "transmuted" dimension. It is the line on the plane.

The object may also bear the marks of the effects of a force and hence make that force manifest in the form of a physical dimension, in the way that the cross-section of a tree describes within just two dimensions both its diameter and the seasons, thus making manifest the measure of time.

As the presence of an object may be described using just a single dimension and the two opposing directions of the unique force which may be applied to it (one transmuted direction each), it can be established that the space required to universally represent any object comprises three dimensions.

As the force is unique no other transmuted dimensions exist. And so it suffices to restore an object with the two physical dimensions necessary for its manifestation to obtain the space for its universal unfolding and manifestation.

This space therefore has five dimensions of which three are native and physical (length, width, height) and two are transmuted dimensions which we are now in a position to name: time, which renders manifest and describes the age of each position of the object, and gravity, which makes manifest and describes the mass of an object.

Given that representation needs like any object to be made manifest, these two spaces share the same system of organisation, the same mechanics of physical dimensions, which are only present within the representation - the same "gnomon".

This universal definition of spaces comprises the three orientations of the three universal physical dimensions making it possible to compare all the objects in the world and, within each set of orientation. The two opposing directions in which the universal force can apply.

And so once the dimensions are gathered together under the aegis of a universal mechanics, it is possible to establish that the space of representation comprises three dimensions and the space of manifestation necessary for our perception comprises five dimensions.

Five-dimensional space is sufficient to contain any object and its representation.

31

1c.1 : An object, whatever its form is included in a space composed of a limited number of dimensions.

1c.2 : Three dimensional space is the only on which can include the representation of any object whatever its volume.

The Uni-verse

We perceive the world as a series images endowed with volume and laid out along the direction in which we are exploring.

Each of them is a snapshot which can then reveal the three native dimensions and the mechanics organising them. Space thus defined is common to all perceptual and representable objects. There is no other space.

Therefore all observers irrespective of their point of view perceive the same evidence. The space of representation is universal.

The representation of this evidence is itself an object in turn and, like all objects in the world, it contains within itself the space necessary for its manifestation.

The extension of these conscious snapshots (that we call our apperceptions) along the direction we are exploring in supposes that the space of representation and the space of manifestation of the world be concentric.

It is the animation consecutive to our passing from one perception to the next that reveals to us the two dimensions distinguishing one space from the other.

Without a space of manifestation there would not be any space of representation, and vice versa. The space of representation is contained within the space of manifestation.

By extending the space of representation and providing it with two supplementary dimensions, the force – as it is unique – establishes it as a space that is universal too.

This space, which thus bears the world and its representation, is what we call the *Universe.*

In one direction the force derives the world from the Universe, and in the other direction it reintegrates it.

The manifestation of objects and perception of their evidence thus constitutes a cycle which though without cause driving all possibilities, an "unmoved mover" as defined by Aristotle.

This "closed cycle engine" is heir to itself and draws its energy from the unique force. It makes all movement possible, together with all the translations and rotations of the slightest pieces of the Universe.

Combined with itself it is the origin of the centrifugal movement of space, like the big Bang, as well as being the source of the centripetal movement of this same space. It alone constitutes the cycle of action and reaction. It is omnipotent and universal.

This "moving force behind the unfolding of the world" cannot oblige the observer to explore the series of snapshots of the world in one direction rather than in another. Nevertheless, thanks to its position at the intersection between the object and the observer, it guarantees that the direction in which apperceptions travel past is always opposed to the direction in which they are explored. Otherwise the observer would always see the same snapshot.

The potential of any image to act as the point of departure for exploring the universe is the corollary of the universal evidence characterising images. Consequently the observer cannot be certain to be presented with the evidence of the same object with each new apperception.

The constraints relating to the unfolding and perception of the world therefore suppose that one do without *the immediate certainty of its existence.*

2. THE ILLUSION OF REALITY

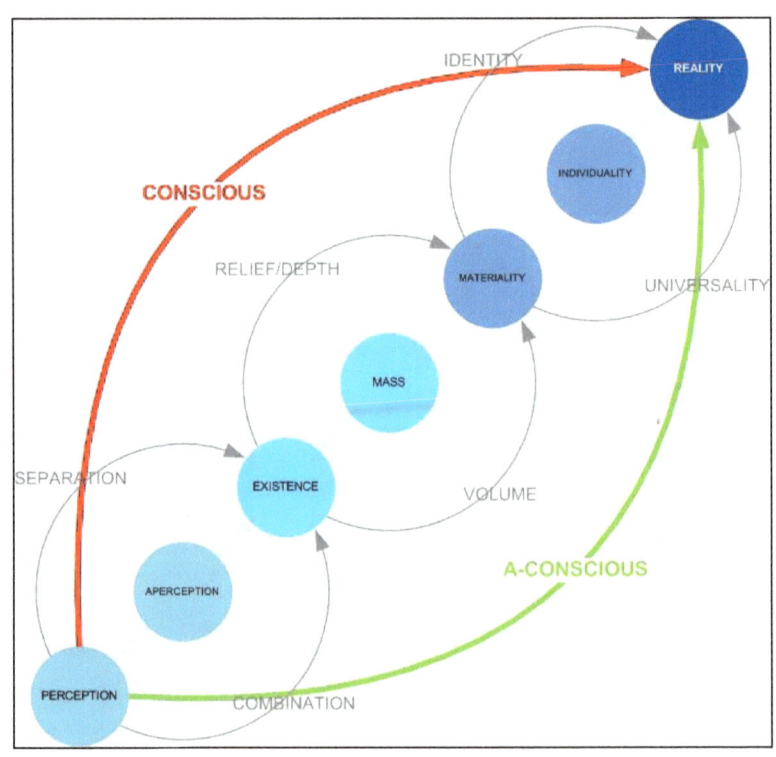

Nothing other than the world is accessible to us.

And we cannot perceive or even conceive of anything outside the limits of the first and unique Universe from which it is derived, without that thing thereby being instantaneously reintegrated within the Universe.

Nothing exists outside the universe or can reach us.

Therefore the instruments with which we endeavour to describe the world can only be within the world.

Given this state of absolute dependency describing the world is, unless we are careful, subject to a fundamental and undetectable illusion. Hence how can we lay claim any certainty about the world and its reality?

As we progress towards the horizon the instruments for describing the world gradually reveal themselves.

Like a mirage in the desert, they only appear and are useful at the place where they reveal themselves. It is illuminating to use them to understand and describe the path taken.

But however great the temptation may be, seeking to use them to extrapolate about the path to be taken - or the world – is to go beyond their capabilities.

At each stage we need to accept to recast our vision of the world.

Given that we are unable to free ourselves of this prime uncertainty, or somehow move beyond it, to ceaselessly try to discover a superior reality beyond the horizon, which would be illusory in turn, it is better to concentrate our efforts elsewhere.

And that is why we shall analyse the process at the very source of this prime relativity or indeterminacy, something which if it is not complete is at least permanent, certain, and universal.

Drawing on the principle that reducing a set to a unit brings us closer to its most efficient structure, let us start by re-examining our world so as to seek its form.

It was in such a way that the astronomers, by giving the solar system its simplest and most coherent mechanics, discovered the first laws of the

Universe and their corollary – that the planets are round. It was not by going around the world.

A universal idea, even if not "proven" by experience or revealed by observation, is nevertheless real, even if initially viewed as erroneous by scientific.

It is by following the example of the astronomers and calling into question our own position as creators and point of reference that our instruments – which are identical to those of our antipodes – will finally enable us to pierce the limits imposed by the veil of illusion, and thereby arrive at an accurate general description of the world.

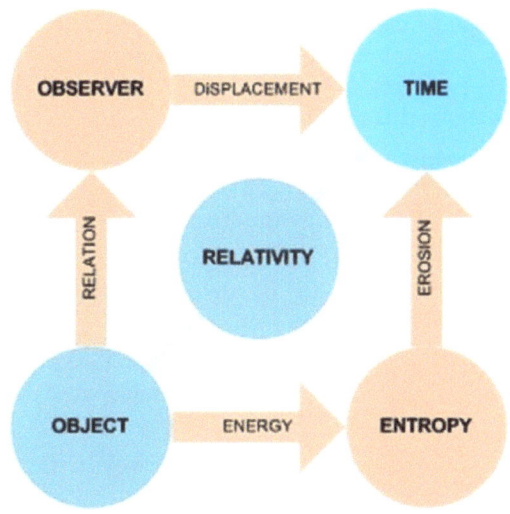

The Relativity of Time: the Event Horizon

Time is not a component of the fundamental structure of the Universe. It is not its fourth dimension.

It is the transmuted form of a dimension of the Universe, which has to be substituted for us to be able to perceive and arrive at a *conscious representation* of the world.

Shorn of any absolute reference this "substituted dimension" cannot be assimilated to the universal dimensions. Time as it is given within the world is therefore irremediably "relative".

Mineral erosion and organic decay are the most current manifestations and references used for the passage of time, what we call entropy.

This is the way we perceive the sand flowing in an hourglass, with the movement consecutive to the loss of structure of a reference object (the upper bulb) to another object which becomes the new reference object (in this instance the lower bowl) or environment.

Defining the standard for measuring time is therefore subsequent to the observation of the movement of a first and indivisible particle - but without yet saying anything about its reality or denomination –, of a unit of matter and energy such as the grain of sand.

We need to recognise that we are only able to perceive this particle because we "encounter" it or at least perceive the energy it is emitting. As of this moment – the moment of observation – both the particle and we ourselves are modified by this encounter.

Since it is visible the particle is divisible, it is thus not spared entropy, and nor is it endowed with motion of such regularity of speed as its supposed unity could lead us to believe. On the other hand its events are certain.

The regular orbits of the planets are another standard yardstick commonly used to measure time. It is true that their great size renders them insensitive to any encounter and, apparently at least, ensures us of their regularity.

Yet the fact that they emit photons regularly until they are reduced to primal unity, without which they would not be visible, destroys these favourable a priori.

On the other hand their characteristics make the flow of time certain and regular, and conversely render events uncertain.

Thus however minute or sizeable these measurement standards may be, they are imperfect and confront us with the "paradox of time".

It is this paradox that Jules Verne illustrates in his novel. His hero, heading east, completes his journey around the world convinced that he is one day behind on his bet.

The judges, who remained in London, are for their part certain that he has succeeded. If he had headed west he would have been convinced he was a day ahead and the judges would have been certain he had failed.

Each one refers to the same "absolute" yardsticks, the cycle of the sun and the bells of Big Ben, and they end up with two contradictory truths.

It would have been the same thing if they had used the "perfect" regularity of the Greenwich atomic clock or any other object with an apparently fixed position, for relative to something else (here the Earth going around the sun) this object is in motion too.

This novel announces Einstein's theory which, barely 40 years later, showed that the speed of movement and altitude modify the flow of time, and consequently our perception of it.

A complex being, especially if endowed with consciousness – let us say man – is subject to regular entropy. And although he is free to move he is affected by the influences deriving from the units and sets which surround him or of which he is composed.

So even if the flow of time and events are known to him and apparently certain, he is unable to free himself of his presence in the world and the measurement standards he finds there, and so arrive at a unit of time. His perception of it is therefore necessary relative.

Finally we have rationally to acknowledge that if we wish to stipulate an absolute reference yardstick for time, it would also have to be unaffected by any force whatsoever or by entropy.

The invisible and motionless galactic abyss, which is omnipresent and contrasting with the stars to which it is the backdrop, could well fulfil this office. But whilst for these reasons it renders the origin of time certain to us, it is not suitable as a means of measurement.

It is at best an incomplete yardstick.

The world does not offer us any other alternative. Since any a yardstick for time has to be visible, it will inherit the world's sensitivity to entropy and to collisions with exterior phenomena.

The time to which the world is present does not fulfil any of the necessary conditions for acceding to the status of dimension of the Universe.

With regard to this dimension we are obliged to make do with a vague origin and imprecise units of measurement provided by natural substitutes and synthetic imitations, which are irremediably relative.

Nevertheless this "imperfect" time assures us of the order of events and the fact that they flow in only one direction. It makes it possible to distinguish them from one another.

And although it cannot be a dimension of the fundamental structure of the Universe, time is a dimension for the manifestation and perception of the world, and as essential a one as the dimensions of space.

Without time nothing could be perceived, everything would disappear.

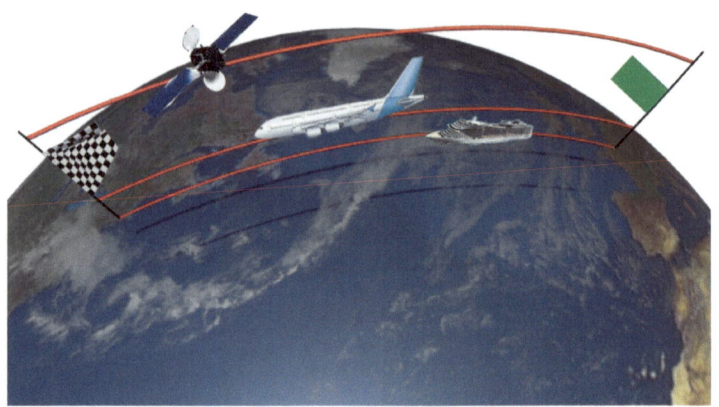

2a.1 : Because of the curvature of the Earth the distance between 2 points is further for a plane or a satellite than for a ship. As so, for the same speed, theirs time of travel would be different.

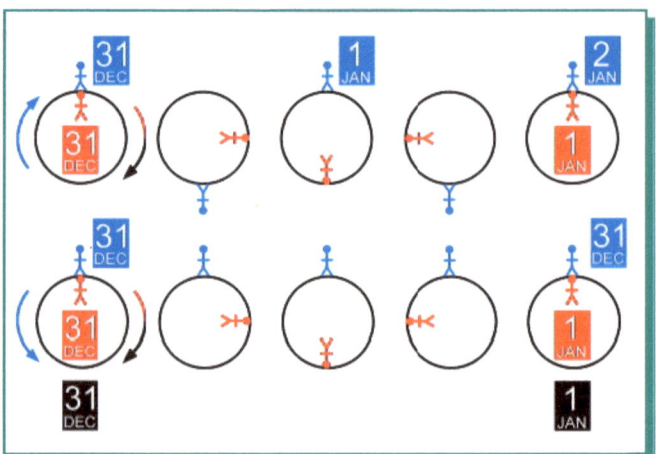

2a.2 : Leaving in the sunrise the traveler (blue) accomplish one more turn than the referee (red) in the time it take the Earth (black) to rotate one time. He then arrive believing it took him two days arrive. Leaving in the sunset the same traveler would turn in the opposite direction of the Earth and then would believe he arrived the same day he departed.

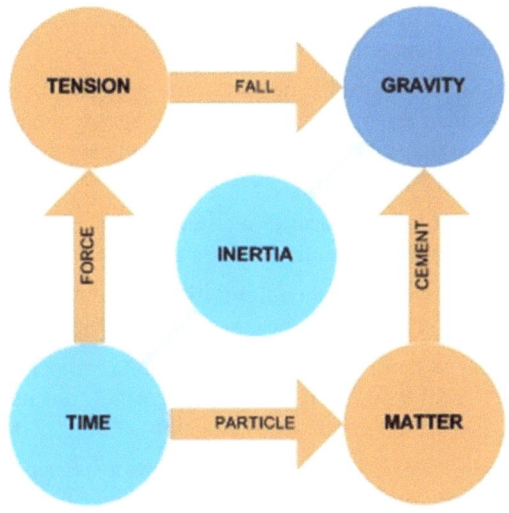

The Hypothesis of Gravity: the Object

Gravity is the name we use for the force that maintains the structure and quantum cohesion of bodies and which, since Newton, Lagrange, and Kepler, enables us to explain the orbits of the planets and stars. It is conversely the same force of gravity that breaks their structures and causes them to fall, collapse or impact. Its effects are omnipresent.

Yet it is still the impassable limit of modern physics. And if, as physicists claim, it is the invisible glue holding together all their hypotheses about matter, it is so elusive and intangible that we still hesitate to consider it as a native dimension of the Universe.

Unlike time and entropy, the fact that an object is attracted towards a star or planet – the principal effect of gravity – has no physical supporting medium other than these two objects. It is relative by nature.

And if we can see that two objects are attracted towards each other, gravity always makes us perceive –paradoxically – the lighter of the two "fall" towards the heavier.

But as shown by lunar exploration, when the astronauts dropped an eagle feather and a hammer from the same height they reached the ground at the same instant, showing that there is no direct and equal relationship between the mass of a falling object and its weight.

Consequently, if we use the same device to measure our mass on the moon it is six times less than it is on earth, and at the Lagrangian point, where the gravitational pull of the Earth and the Moon cancel each other out, it would be absolutely zero.

This equilibrium enables the Earth to retain its satellite as well as enabling objects, depending upon their state, to retain their shape and volume.

Thus an uncontained gas retains neither its volume nor any form. The molecules composing it – of water for instance – do not form any structure.

A liquid on the other hand – made up of the same molecules – retains its volume irrespective of whether it is contained or not, but its shape is only that of the container. If not contained it is a blob on a supporting medium.

In a state of weightlessness this same liquid forms a spherical bubble, illustrating the reciprocal and equal attraction maintaining the relative position of the molecules within its structure.

A piece of ice resulting from the solidification of this liquid has a structure of even greater equilibrium. It inherits the liquid's ability to retain its volume, but is further able to retain its form. Thus two ice cubes of a different nature do not mix.

Once combined and decanted into a glass, the "lighter" liquid appears to float on the "heavier" liquid, such as oil on water for example.

These two volumes are distinguished from each other by the difference in the ratio between their mass and their volume, what is commonly called their density, or more accurately their mass density.

It is this phenomenon which explains the difference in attraction of the Earth and the Moon, and which enables black holes, where it is infinite in value, to attract all forms of structure and energy, including light.

Given that gravity only exists as the form of attraction between two (infinitely small or infinitely large) objects, it affects and originates in all the objects in the world, and has as many units of measurement as there are objects in the world.

And if the galactic abyss is the origin of time, it is the full set of celestial bodies which constitutes the (equally diffuse) origin of the overall gravity of the world.

Yet if, following on from Einstein, we accept that a photon, a corpuscle of light (in the classical physical scheme) has a constant speed despite the fact that time is relative, then if we are to be coherent we have to accept the existence of a vehicle, as a corollary to gravity, whose speed is also constant but whose direction is always diametrically opposed to that of the photon.

If it met these conditions it would, paradoxically, be definitively invisible.

We may deduce that the energy exerted by the attraction of the black hole and that exerted by the light to free itself cancel each other out at the event horizon of a black hole, in a comparable state of equilibrium to that found at the Lagrangian points between the attraction of two heavenly bodies, or that found on the rim of a glass between gravity and the surface tension of the liquid.

The two corpuscles meet and become one, time being equal to gravity.

We are unable to conceive of a unit of measurement for gravity without being able perceive its effects, and we are equally unable to establish a measurement standard for it as we were for time.

Gravity is as relative as time, and like time it has no claim on being a dimension of the Universe.

Nevertheless gravity and its effects are a necessary condition for the manifestation of the world and its perception. The way the structure of all objects is constituted is due to gravity, as is the fact that they differ from one another in form and in mass density.

Gravity is the source of attraction between objects and the origin of impacts instrumental to their disintegration, and is both a complement to time and its opposite. Whilst time makes it possible to separate events, gravity separate states and forms, rendering them distinct.

Gravity is a dimension of the world in the same way as space and time. It is the transmuted form of a fundamental dimension of the Universe.

Without gravity nothing exists, everything collapses.

*2b.1 : Because of the difference in gravity, the golf ball (of constant mass) will fly far-
ther on the Moon than on the Earth, and in definitively in space.*

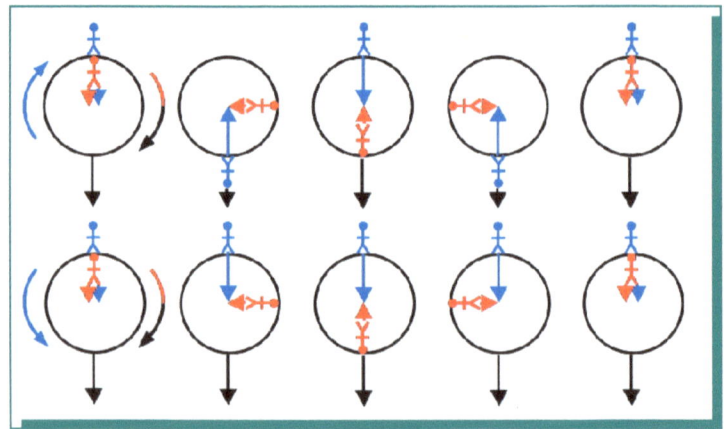

*2b.2 : The pull of gravity set by the sun on the earth and the one set by the earth on
both the traveler (blue) and the referee (red) are of the same direction at both the start
and finish of the race. Only when the traveler set sail in the sunset is he keeping both
pull of gravity in the same direction all along his journey.*

46

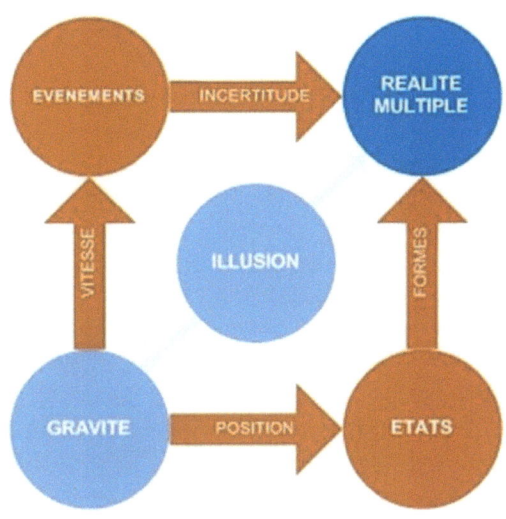

A Multiple Reality: Indeterminacy

Time and gravity are the elements of our world that ensure the one-directional order of events and states. It is their combination which gives rise to the irreversibility in the way events and states are linked up, what we call *causality*.

The "laws" which derive from this form the postulate by which we distinguish the real from the improbable in a way that is universally evident.

Their joint relativity paradoxically means it is not possible to specify the form or emerging state of an event. At the very most causality means that link between events and states is something of which can be certain, without them necessarily occurring or even existing. This relativity means we are confronted with a multiple and indeterminate reality.

Hence the blob we discover on a supporting medium is necessarily the consequence of the impact between a drop of liquid and an obstacle. The laws of entropy even enable us to be certain of the relative dimensions of the drop from which the blob is issued.

But as we did not perceive the initial event we have to acknowledge that we know nothing at all about the moment proceeding impact, though with-

47

out calling its existence into doubt, even if it is highly likely that the drop fell "from above" due to gravity.

If on the other hand we perceive the drop of water in a state of suspension, it is certain due once again to causality that the drop will collide under the effect of gravity with an obstacle, and consecutively change state to form on some supporting medium a blob whose dimensions are certain but whose shape is uncertain.

The relativity of the elements of causality confronts us with one of reality's mysteries that Schrödinger illustrated in his famous thought experiment in which a cat, a flask of poison, and a hammer are placed in a sealed box. It is certain that the cat will die at some stage but not when it will die.

For as long as the lid of the box remains closed the inside and the outside (with which the observer is associated) belonged to distinct realities. As one contains the other we may speak here of "levels of reality".

The contents of the box (the cat, flask, and hammer) are invisible, and so the observer cannot confirm the existence of these contents nor determine their state. Nor may he impose "his" causality, on the contents. The level of reality of the contents is free to follow another postulate.

The cat can thus die, resuscitate, or else remain in one or the other of these states. It is only when the lid is taken off and the two levels of reality are reunited and submitted to the same postulate of causality that this set becomes perceptible and tangible, hence made real and certain.

The death of the cat is only certain when constrained to the same level of reality as the observer and the closed box, or else once the box has been opened, when only a single level of reality remains.

If we perceive the world and all the objects of which it is composed it is because, under the effect of entropy and the impacts in which they have been involved, they emit, reflect, or refract the energy and particles that we then encounter, thus revealing their existence to us.

Consequently we cannot perceive an object emitting no energy or attracting no energy – and hence we cannot consider any such object as being part of our level of reality.

Yet we can distinguish black holes as they contrast with planets or stars. Hence as a matter of complementarity there exists a level of reality founded on a postulate of causality opposed to that founding our level of reality.

An observer can only be present to entropy working in one direction, but without being limited to experiencing only this postulate.

If we modify the direction of events relative to causality (by playing a film backwards, for example) it is possible for the same observer to admit that two states or opposing events are equally real, provided he perceives only one of them.

Gravity can thus cause a glass to break and shatter, or it can cause it to rise up and be reassembled on the edge of a table, it all depends on the postulate according to which the observer is present.

If the observer is confronted with both at the same time he will see neither. If he perceives one following on from the other, then the first will be considered as the norm, thus removing any ambiguity with regard to the chronological direction of time and gravity, and so defining reality. The second becomes illusion.

The only further criteria to this distinction is the a priori instituted by the observer himself, originating in his postulate of causality. As a consequence of this relation the reality which is distinguished and perceived by an observer can only lead to uncertain and incomplete laws.

Drawing on the same principle of complementarity as that which enables an observer to conceive of a black hole, everything that he cannot directly perceive in the world – and by extension the Universe – is hence not constrained to operate according to the same postulate of causality as that applying where he is present.

The Universe is not uniquely and definitively constituted solely of the levels of reality founded on the same postulate of causality as that of the observer.

Limiting the Universe in such a way would involve conflating world and Universe.

This confusion illustrates the universal and comforting misconception with regard to the relativity of time and gravity. This misunderstanding is born of the need to perceive the world as the real face of an uncertain Universe.

This reality is the least uncertain.

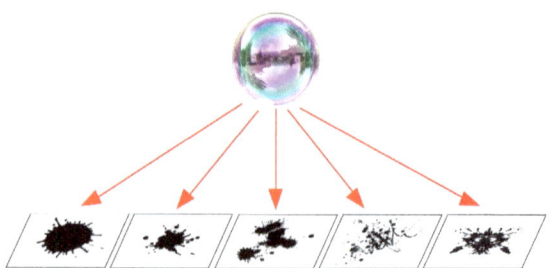

2c.1 : Gravity is the cause of every stain, which would be different from one another whatever the number of stain.

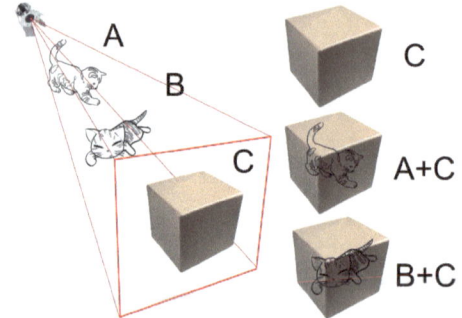

2c.2 : There are as many realities as there are combinations of elements

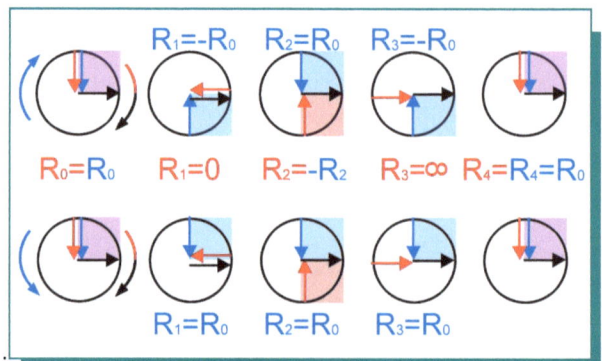

2c.3 : The reality set of both the referee and the traveler is defined by the direction of time (black arrow) and the direction of gravity each one perceive. Both realities are the same when both of them occupy the same position. When one of them occupies a position at other side of the earth, its perceived gravity is of a direction opposite the one he perceived earlier, consequently so for his reality set.

50

Manifestation and representation

Evident appearance exists as much as reality does.

It is only when we cross (in one direction or the other) the screen of the event horizon on which the evidence of the world appears to us (as we pass "behind the curtain") that we can distinguish mirage from reality.

What each of us perceives of the world is in fact entirely illusory. It is an illusion itself born of another illusion. The trick certifying the reality of what we perceive and the realism of what we imagine is causality.

The perfect example of this is the impossibility of distinguishing between a computer-generated image and a true image, both of which appear equally "virtual" on the screen on which a film is shown. The absence of any discriminating contrasts may be explained – above and beyond our acquired knowledge – by the relativity of the reference standards on which the reality of the world is based, time and gravity.

The centrifugal movement consecutive to the entropy of objects is the driving force behind our perception of the world. It is this which "unrolls" time. Conversely, the centripetal movement consecutive to the attraction of elements and objects constitutes the driving force behind the manifestation of the world.

It is the combination of these two opposites which certifies what is in fact mere illusion as being evident and real, and this consequently enables us to call into question the reality of what it certifies as true. It is the mechanics according to which the world "becomes real".

Given that we are convinced the universe is immobile despite the Hubble constant demonstrating its state of expansion, the components of this mechanics - though relative and manipulable - place us in the position of conceiving that universal attraction is "rolling up" the Universe.

We ought to conclude that the foundations underpinning the evident reality of our world, the foundations constitutive of causality, are in fact artifices serving the manifestation of the world.

If the reflection of the moon appears to be identical to everyone, a cone of light facing in the direction of the observer, the only thing he may reasonably consider to be real is the reflection that he perceives. The reflections appearing with an equal degree of evidence to other observers are, for him, artefacts.

But unless we accept that the entire ocean is covered with the reflection of the moon, we have to recognise that these reflections are a collective illusion.

The image of the evidence of the world is the product of its manifestation and its perception by the observer, both of which are placed on opposite side of the screen. Without any manifestation of the world, the world could not be perceived; and without anybody to perceive it, it could not be made manifest.

The evidence of the world is a mere hologram that is present at our event horizon, the product of two projections each with a distinct source but guided by the same reversible causality. The set of these two opposing occurrences of causality is for its part irreversible.

The evident reality of the world is the result of a double misapprehension.

Consequently the fundamentals laws of the Universe *cannot include time or gravity in their equations and need to be free of all causality.*

3. THE CONSTRUCTION OF THE UNIVERSE

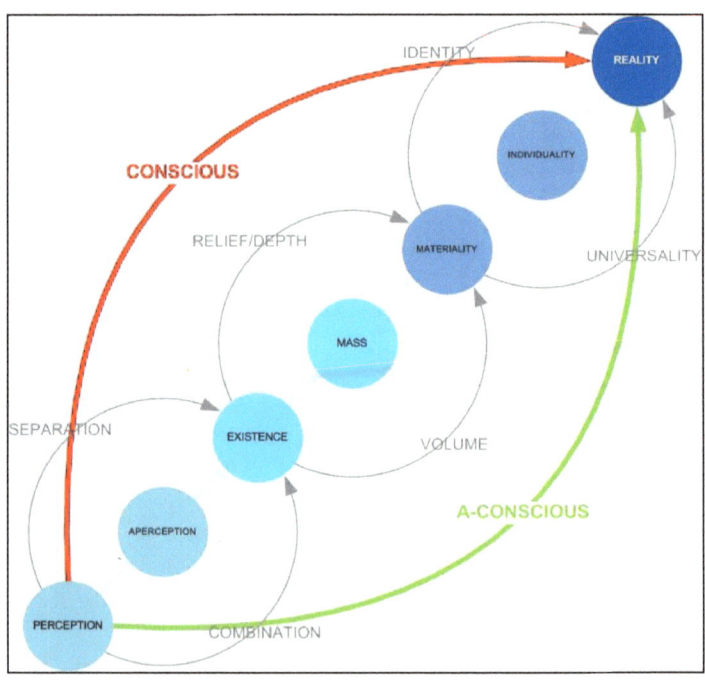

Our experience of the world is (like the evidence of the world) unable to banish our doubts about its perfect materiality, and even less able to provide us with certain laws relating to its operation – given that the world, as soon as we wake up and leave behind the space of dreams, imposes its omnipresence which is merely the illusion we call reality.

Yet it is on the basis of the components of this same elusive and intangible presence, and by exploring the same illusory world, that we first build up our conception and then our representation of a Universe as we go around experiencing it The Universe is the only place of certainty, from which our perceptible world is derived whilst being at the same time part of it.

The perceptible world unfolds within the space emerging from the combination of dimensions and forces, and is made manifest by the operation of time and gravity.

This is instrumental in our sharing a conscious representation of the same world and Universe, independently of our individual solitary experiences. But without thereby guaranteeing anything further,

Nevertheless prior even to awareness we are "present" to a second instance of the world – an a-conscious instance, the space of which is made up by imagination and dream *inter alia*.

Although similar to our waking world for it is peopled by the perceptions we have experienced, there are no spatial or temporal constraints limiting its field. All of the Universe may be explored within it and each event may be experienced infinitely via a process of mysterious inspiration.

Everything which is invisible or unforeseeable may be perceived within it, everything which is unknown may be present, and everything which is impossible is possible there.

This world we dream of as perfect (in its beauty or horror) appears to us in the same terms as the world we consciously perceive when awake. It is real in an unreal way.

Although sight, hearing, smell, taste, and touch enable us to apprehend a specific *part* of these two worlds, only sight and hearing have a mechanics that can help discover their *structure*.

The absence of any medium being able to convey an emission to a receiver and to make that emission persist in time tells us with which sense we could perceive a world without time, as if it were freeze-framed. It is sight alone which is capable of such a feat and able to constitute an immediate and instantaneous representation.

We shall therefore accord it priority here – as is also the case in the adventures of our dreams – without any further justification.

That being said, the structure common to all dreamt possibilities only becomes apparent to us once we are awake, once we are present to the so-called real world, and so called to differentiate it from the world of dreams.

Given that the counterpart of its "reality" is the omnipresence of causality and the limits imposed by time and space (unlike the world we inhabit during sleep), this world can only be apprehended partially, in bits.

As we reassemble the jigsaw puzzle of our perceptions, we discover the full perfection of the other world - though without ever attaining any such perfection.

It is by exploring this process of conscious perception of the world that we will be able to understand what the structure of this second representation is, where everything is possible.

Once we have done this we will be able to build a model of the Universe which will act as a matrix for all possibilities within all possible worlds (including our world), *both the origin and the compendium of all Laws.*

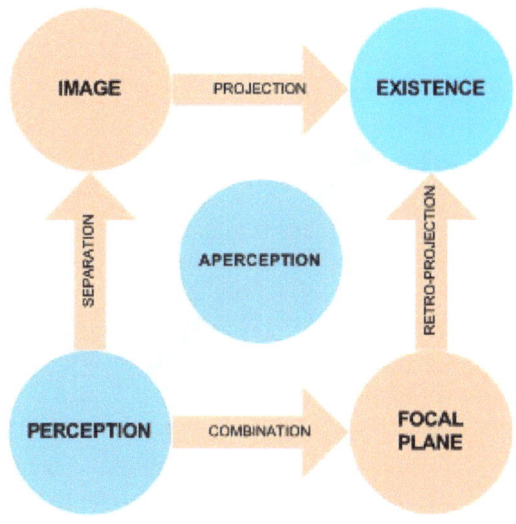

The Image of the Parts

Whenever we are awake to the world our first perceptive contact, our first *perception* of the world takes place outside time or, rather, as if it were "placed" within a primal, motionless time.

Our acquisition of the *existence* of the world is accompanied by our perception of snapshots of only those parts of the world which our perceptive organs are capable of detecting. The world never offers itself to us in its totality, in its infinite extent, or in its infinite detail.

But nor does it offer itself to us with nothing. Our perceptive system is so constituted that it can only perceive what it can "distinguish", by which we mean separate a part from a whole, a part that we will here call an object.

The *image* of the world that derives from this initial perception, along with all the following perceptions, is only present within the ether separating the sensing subject from the projecting object.

Its only supporting medium is the source of its projection and the sensor detecting it. But there is not the space necessary for it to be made manifest or to unfold in these two points.

57

If this snapshot of partial perception of the world were one of hearing instead of sight, the product would at best be an audio feedback (or Larsen effect), but it would more likely be a silent sound.

Given that the physical specificities of a sense cannot enter into consideration here, this receptor needs to be seen as a plain, unique, and impressionable surface.

The initial image of the world thus acquired is fixed, unique, and double (for reasons similar to why the world is a double illusion). It is in fact the combination of the *direct projection* from the object or world and the *rear projection* of the same perception but from our own senses.

This "rear projection" of the image is absolutely necessary for perception to occur. A monocular sensing subject or "Cyclops" such as a camera operator can only take a sharp image projected by an object on the *focal plane*.

And he can only observe this image once it is projected in turn onto a screen or transferred to a paper medium, on a second focal plane. And so there are two projections which are wholly identical other than in terms of their source.

This being said, whilst we do not call into question the evidence and reality of the world on acquiring this initial image, this is not the case for the "parts" of objects of which the world is composed. Within this illusory image the object must be distinguished from the rest if it is to exist. We need to be able to perceive the world and the object.

And a single image does not suffice to do this.

Each image unfolds in just two dimensions, to which are associated the dimension across the ether between the projector and the receiver – that of *primordial time,* the time of psychic processing of perception, or apperception.

No one sensing subject (a "Cyclops" once again) among all the sensing subjects present to the object has any chance on its own of distinguishing it from the world, whether by modifying its *position* around the object or altering the object's *attitude*. The time of apperception does not make this possible.

But if they *combine* their perceptions the image produced by their rear projections will appear as if issuing from a single observer. Since it is not possible for several sensors to occupy the same position at the same time, the perceived images conflict with each other.

58

In the time of apperception this situation cannot result in distinguishing an object.

But if on the contrary a single sensing subject combines the perceptions from at least two separate receptors, he would in the time of apperception be able to obtain a unique image where each perception acts mutually as a referent.

Like a coincidence rangefinder, the parallax resulting from the necessarily different situation of the two receptors in relation to the object is precisely what makes it possible to supplement the two dimensions of space of each perception, which are sufficient for the object to be *present* to the world, with a third dimension specifying the distance between it and the observer and giving the illusion that the object has relief.

Without going into the details of the neuroanatomical or psychic specifics involved in the processing of the image by the human brain (in particular the double-half-image detected by each eye), suffice it to say that the sensing subject is equipped with stereoscopic vision.

Although it is fixed, the unified, rear-projected image obtained in this way is – either wholly or partially – "out of focus", like on an incorrectly adjusted camera. The fact that it is out of focus is a consequence of the combination of the two occurrences of (nascent) time necessary for monocular apperception. It is the derived representation of Time and thus what makes it visible.

The stereoscopic acquisition of this immaterial (as without mass) and instantaneous image therefore reinstates (or gives?) all its dimensions of space to the world, as well as bringing a new dimension which acts as the "cradle" of Time.

Endowed with all these dimensions the world and object each acquire their *"ex-istence"*.

However even when the image issuing from this process of apperception extends in all of its dimensions, it never represents more than a part of this world within which objects only ever present the illusion of being distinct from it.

But it already carries within it all the keys for *building the Universe*.

3a.1 : Invisible at first sight, the gangway appear only when one change position.

 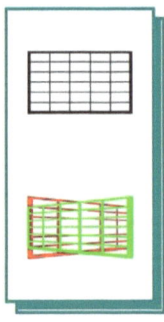

3a.2 : Stereo photography camera and principle of coincidence perspective.

60

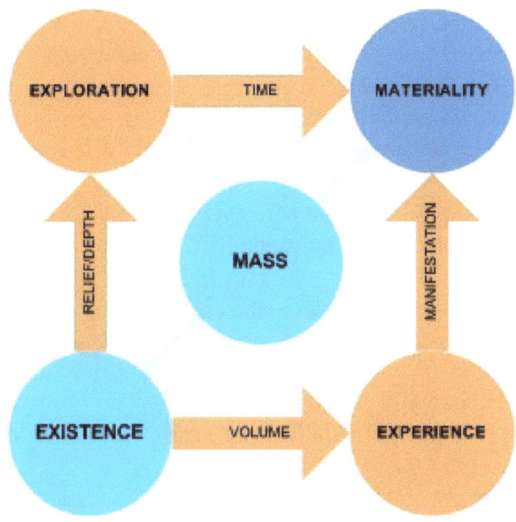

The Space-Time Curve

The illusion created by the relief affecting the image and issuing from initial apperception, and the fact that it is out of focus, incite the observer to push his exploration further so as to determine whether the volume of the object belongs to the world or is distinct from it, and hence possessing its own substance or *materiality*.

This presupposes that there be a limited and privileged relationship in which the object and the observer – and only these two – are made manifest to each other.

It also presupposes that there be a reference dimension enabling the observer to organise and compare his perceptions following the order of the positions he has successively occupied. This enables the observer to build up a conscious representation of the *volume* of the object by following this kinetics. Otherwise a third party is required to produce a complete experience of a single object.

The reference dimension that is doubly necessary for the acquisition of knowledge of an object and, by extrapolation, of the world and Universe, is Time, resulting from the primordial time of apperception.

Because of this association between space and time the surface covered by an image issuing from each apperception is limited in space and time, above and beyond the resolution specific to each sensor. It becomes impossible for the observer to explore space without equally exploring time.

This reference dimension makes it possible even for a Cyclops to reconstitute, thanks to back-ups, an identical image to that seen by a stereoscopic observer, by combining two consecutive perceptions at any moment of its experience.

Without this duplication all of the perceived images of the object could only constitute a discontinuous representation, a jigsaw puzzle without any edges, cut-out pieces, or model.

The plot of this dimension represents the cumulative state of *exploration* of the volume of the object by each observer, what we call their *experiential trajectory*. This trajectory enables the observer not only to experience the object but also incidentally experience "others" (other observers) due to the fact that other observers potentially occupy different positions around that object.

And if we all experience the same object we have to acknowledge that we all share the same reference dimension. The trajectories taken by the observers are necessarily different, and so they are all plots of the unique and specific occurrences of each observer on the reference dimension.

The reference dimension therefore constitutes the basis of a universal structure linking up all our experiences and all our perceptions, but which is nevertheless absent. Our experiential trajectory is the derivative representation of our movement on this intangible and elusive structure.

Without this shared system of space and time each person would only be present to his own personal world, of which he would be the centre and unique inhabitant. Other individuals would be as inaccessible to him as their worlds. There would be no other Universe. The world and Universe would coincide. In addition to binding object and observer and more generally space and time into a relation of restricted relativity, this system continues to cause the observer's conscious representation of the object to be subject to motion blur, which may loosely speaking be said to be due to movement.

The ambivalence and equivalence between the object and the observer mean it is not possible to establish with any certainty which is experiencing which. It is equally impossible to establish which is in motion and which is at rest without including a third party (or referent) within this restricted relationship. They are thus both the source of this lack of focus.

The lack of focus perceived by the observer is indicative of the time concomitant to his own movement and of the time concomitant to the movement of the object. The concomitant movement and time of the two, being situated at the opposite ends of the object-observer set, are diametrically opposed.

If the lack of focus of the "observer" is always considered as positive, the lack of focus of the "object" is thereby negative, as are its movement and concomitant time. As neither can relive both the position and instant they have just left, the object must be admitted to be immobile. The lack of focus of the "object" thus represents its *inertia*, what keeps it in place. It is indicative of its substance, what we call its *mass*.

This lack of focus is the manifestation of the inertia of the object-observer set, as well as manifesting the inertia of the space-time pair.

Each perception projected onto a focal plane is by its very nature in two dimensions and perfectly sharp, with the blur being the consequence of the deformation the images undergo – without the intervention of any (optical) device – so as to conform with the common a-conscious structure and, thus placed in "perspective", be added to and combine with contiguous perceptions.

From the point of view of the observer the rear-projected image situated between him and the object undergoes an analogous deformation in conforming to the structure of the object to that which is consecutive to its being perceived through a "mental" convex lens whose focal point is the object. But to conform to the structure of a representation the image undergoes a concave deformation whose focal point is the observer. During this one-directional trajectory between the two, each perceived image is deformed twice in a perfectly symmetrical manner. At the point tangential to these two deformations lies the focal plane where perceived images *appear* to us.

At this point the mass deforming the manifestation of the object, and time that deforms our representation, are in a state of perfect equilibrium. It is at this particular place in the structure of the object-observer system, which is analogous to a *Lagrangian point*, that the object encounters the space that is common to the universal manifestation of its presence and to our exploration of its volume.

The object, which exists solely due to apperception, acquires *materiality* thanks to the contribution of mass.

3b.1 : Phenakistoscope (left) and Zoetrope (right) are two instruments that give the illusion of movement using the principle of animated images. Only the orientations of their axis of rotation differentiate one from the other.

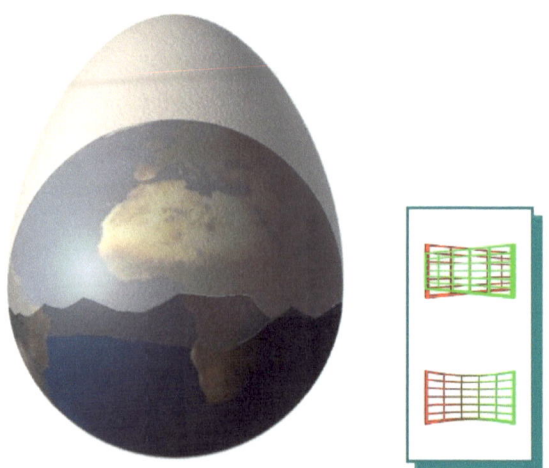

3b.2 : Perception of parts of an object does not reveal all of its shape. Our mind does however project the simplest volume (a primitive) when the object is in fact egg shape.

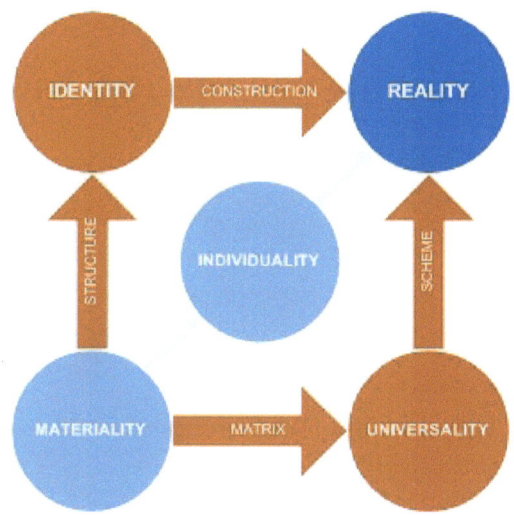

The Sphere of Totality

The Observer no longer has anything to discover about the object that he has already explored from all angles as it were, and so he can now produce a mental representation of the product of this experience by gathering together the scattered perceptions of this object into a single whole, thereby reuniting the All within one.

In order to "geo-locate" these images a log of these *mental co-ordinates* has to be built up, thus making it possible to locate each of them in time and space by following the experiential trajectory.

This plot functions as the pattern (what we will call the *model*) on which the perceived images will be projected and from which the non-perceived parts will be extrapolated in order to reconstitute the object, as well as being the set of possible movements for the observer (who is ultimately only one observer among others) around that object.

This diagram is what we call its *kinetics*.

Because of the linear chain of perceptions deriving from what was the "perfect" reference dimension of time, we all have the illusion that our experiential trajectory is perfectly linear too - like a man lost in the deserts who is actually going round in circles.

Because of the community of reference dimensions it is impossible to conceive of our experiential trajectories as being strictly parallel. They necessarily intersect and each observer has to cross his own trajectory at least once when going around the object.

Thus all the observers receive confirmation that they have experienced the same object. And the observer who *distinguished* it may finally *detach* it from the world. This process of building up a representation from each part to the whole provides the model for our experience of the world.

Though it acts as the model for our experience of the world, it is not the model of our experience of the Universe as it is impossible for any observer to directly experience himself. At best he can acquire knowledge of all of the objects and of the world.

Each projection and each consecutive rear-projection of a perception of the world originates in a common and unique point.

The deformation required for them to combine and build up a *single image* implies that all of the points along the plot be "distended" so as to join up with the contiguous points representing the concomitant consecutive perceptions.

They now form an interrupted line (with a beginning and an end) of the "surfaces" equalling the perceived surface of the object at this point in our experienced of it. The perimeter of each surface corresponds to the event horizon at the moment of perception and constitutes the fundamental unit of the structure of representation.

The image undergoes two consecutive, opposite, and complementary deformations as it goes from the original projecting object to the observer acting as the source of its re-projection. And thus freed of the disturbance of time and gravity in turn, the representation is identical to the object and universal.

When applied to the perceived entirety of the object and to each of the distinct units on its surface, the universal characteristics of these two deformations are not just complementary but also reveal to us the form of the experiential matrix of the world and the volume of the model.

The dual obligation (firstly of seeing the experiential trajectory cross itself at least once, and secondly for it to be extended equally in all directions for each point) can only result in one figure – a *sphere*.

A sphere does not directly restitute the reality of the object, but it does strictly respect its proportions without its spherical form being altered in any way, and so is the perfect structural matrix for the object.

Each distinguished part of an object will be an instance of this sphere.

The set of instances required to cover the totality of the surface of each object combines in turn to form a sphere including them all.

Lastly, all the "object" spheres are contained within the sphere of the world from which these objects have been distinguished, as their parts have been too, in a state of strict and perfect homothety between neighbours and parents.

Furthermore at the very beginning of each observer's experience of the world all of these occurrences are geometrically indistinct. This structure may thus be said to be holographic, permanent, and universal.

Given this it is hardly surprising that all the components of the Universe be perceived – or else are fictively reconstituted – according to the same architecture, be they the smallest elementary particle, atom, or molecule, or the largest heavenly body.

Whilst this structure is not a representation of the object, and even less an image of it, it necessarily partakes in its constitution and so we may conclude that it is a-conscious.

Once we have acquired full knowledge of an object by perceiving it, experiencing it, and then representing it, it thereby becomes a unique object which can no longer be confused with any similar object.

Once its characteristics are reconstituted within its conscious representation they provide the context for its unique occurrence of the universal a-conscious structure, its *identity*.

An object whose image has been distinguished becomes detached from the world and acquires its existence, then it acquires the mass necessary for materiality with the help of the perceptions deriving from the observer's experimentation with it. The sphere of its totality endows it with its individuality.

And so it has thereby secured its *reality*.

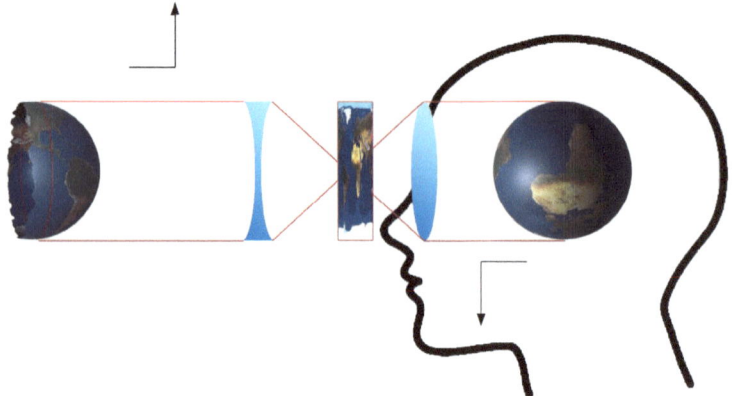

3c.1 : Complementary and consecutive deformations applied to perceived images allow for the construction of an accurate rendering of the whole world.

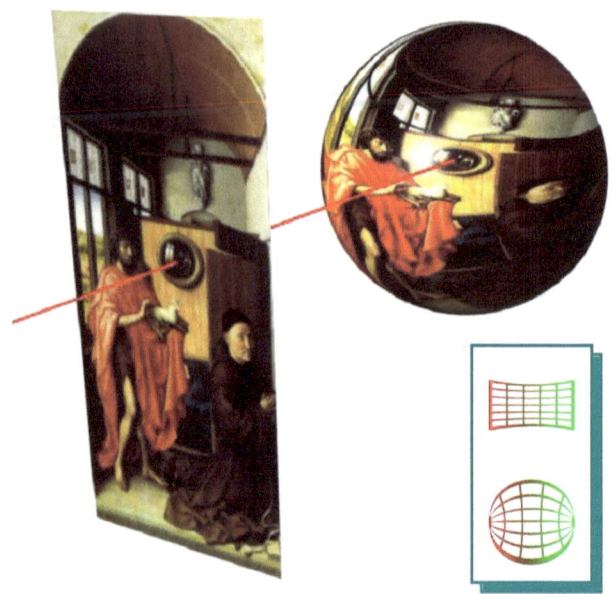

3c.2 : Spherical retro projection put every observer at the center of the world representation. It put him back inside

68

Rediscovered Certainty

Each of our perceptions, be they of the most banal and everyday kind or scientific investigations employing the most sophisticated strategies and apparatus, is an occurrence of the spatial and temporal unity of our experience of the world. It makes up the elusive present.

This experience of the world, divided up between day and night, waking and sleeping, attentiveness and inattentiveness, absence or presence to a specific object, would be unable to partake in a unique and coherent conscious representation were it not for a universal, imperceptible, and a-conscious structure enabling the days and perceptions to follow on from each other without giving the impression of eternally beginning anew.

And were it not for this intuitive structure this experience, divided up into position A and position B, would not be able to combine the perceptions issued from these two positions and so rear-project the image of a complete and unique object.

This structure "contains" all the objects in the world due to its universal and impersonal form, unlike our conscious representation which accounts only for the partial, unique, and subjective nature of our experiential trajectory.

This structure is made up of only those dimensions of space, time, and mass necessary for the manifestation of these objects, which are thus present, in the form of their reflection within our conscious representation. The force driving our representation is consequently inverse to the force driving manifestation – it is its mirror.

The perfect complementarity of these two driving forces - one enabling the world to become manifest and the other enabling us to perceive it the same terms – form a whole which is paradoxically immobile. This mechanics is that of the Universe.

Transmuting matter into energy and then performing the opposite operation, it makes the a-conscious structure the primordial matrix for information. It thereby ensures that this structure, which establishes the continuity of our experience of the world, lasts over time. It is the driving force behind the construction of the Universe.

 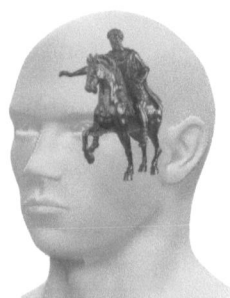

Despite the limits imposed on the details, space, and time of each perception, and despite the discontinuity attendant upon our own experience of the world, our representation inherits this reference matrix's full continuity.

Because of this equal and universal inheritance every individual is a part of the world and the conscious representation which it forms – a part of knowledge of its totality.

Conscious representation is the *Intellect Agent*.

By providing the requisite continuity for the architecture of our representation of the world, this structure retroactively reconstitutes time in the form of the succession of days and nights, inspiring us with the *certainty* that we are not living through a never-ending day.

In conclusion, by delivering mass, it *certifies* the permanent and unique character of each object that we experience and which strictly speaking it is only an instantaneous dismemberment of the All, like the horse freed by Michelangelo from its marble gangue.

The a-conscious representation of Universe *builds up the reality of the world*.

CONCLUSION

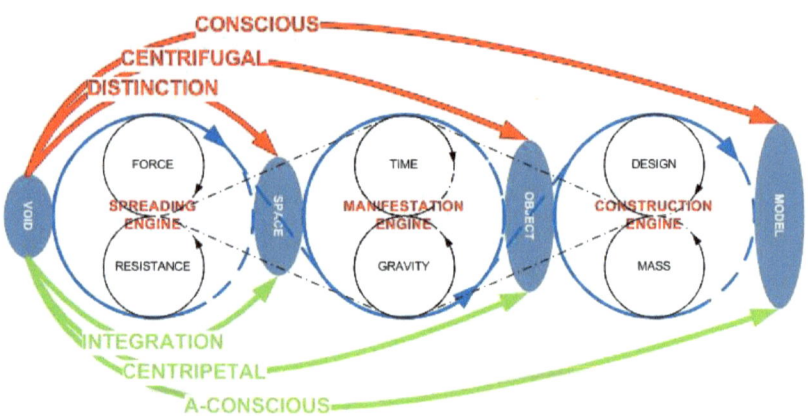

Evidence and reality

Whatever its universality, the evidence of the world that is displayed to us and that we denominate its reality to us remains definitively relative, infinitely uncertain and irremediably incomplete.

There exists not more than one universe, as we might believe, but more than one 'aspect' (or more than one mode) of the same Universe.

This reality to which we are so attached is nothing more than the image derived (through the intervention of time and gravity) of its perfect and absolute reality.

Furthermore, it is an essential condition of our perception and conceptualization of the world – its apperception – that we are only able to embrace a single *aspect* of the Universe at any one time.

Any spontaneous or considered observation – and consequently our entire knowledge of the world – is therefore linked to our relationship to it, what we call our presence in the world. For this reason, the conscious representation that we can make of it can only be definitively relative.

Our experience of the world, therefore subject to the relativity of time and gravity, only enables us to perceive one state or event at a time. The representation that we form of it, in the absence of an overall view of its evolution, is therefore irremediably incomplete.

Also, however sophisticated the means of observation and deduction employed by science to reconstruct the hidden face of the world and thus to reinstate the unity of the All, they only allow us a vague approach due to the impossibility of experiencing it ourselves, thereby keeping us in a state of uncertainty.

The forms of our experience of the world, however relative it may be, and our spontaneous representation of it, though incomplete and uncertain, nevertheless provide our intelligence with a few clues to its *universal structure*.

The Universe is the Matrix of the Worlds

The search for a "world system" governed by certain laws, which has been at the heart of the scientific approach from Galileo down to Newton, Laplace, and modern physics, is by its very nature destined to fail, like the ancient Tower of Babel.

We now know that its frontiers are expanding, its masses are in motion, and its elements are subject to constant entropy (for matter) and constant evolution (for living things).

No architecture can resist this contradiction between a tendency towards ever greater disorder and a tendency towards the most complex order. Science has nevertheless managed to display its own limits – all of its theories will remain "limited".

Nevertheless it is still possible and legitimate to search for a coherent set of principals accounting in a fixed and certain way for the relationship between the All and One, uniting without frontier or demarcation the infinitely large and the infinitely small (the α and the ω) in a space without any "edges", that is to say one that is infinitely finite.

The world is neither chaos nor perfect order, but a subset of a superior order which alone can account for its frontiers, its evolution, and the laws governing it – that superior order is the Universe.

The Universe is the matrix of all possible worlds and in particular the one made manifest to us.

All evolution and entropy must start from an original state. Each movement of mass must come from an emitter and there is no frontier distinguishing between presence and nothingness.

The world cannot act as the theatre for the reconciliation of time and gravity, of order and disorder. So the only thing left to act as the foundation for our representation of the world and of space-time is the Universe. Though it cannot be apprehended by consciousness, the Universe is nevertheless the *absolute* origin.

74

States/Events Duality and Double Representation

All states and events constitute the irreconcilable pairing of the ways in which we experience the world. The evident reality of the one prevents the precise reality of the other being determined, and this despite its evidence. Yet it is still possible to establish a joint representation reuniting them.

Any state made manifest in the world is not a spontaneous appearance of order from disorder but directly related to another state. The manifestation of this link (which we call causality) constitutes the event. Each event therefore has two opposing levels of reality – the father and the son, birth and death.

And the further we move away from the immediate evidence of the event or state to the point of finally considering our world as a whole, the more uncertain it becomes that we can represent, as in Zeno's paradox, the arrow as both reaching the target and having reached it.

Only a double representation is able to make certain what appears more and more uncertain at each level of reality.

The world may be seen as the conjunction of two representations such as those in Schrödinger's intuition. A first representation of perfectly definite form, representing all levels of reality, all the states under consideration irrespective of their number or value.

And a second representation of indefinite form but perfectly circumscribing the first, and necessary so that the event "the cat will die" and "the arrow will reach its target" be certain, together with the manifested states delimiting these events.

Irrespective of the aporetic trajectory taken up by the volume of the world, the inspiration that the event is the space in which these states have to occur and that is only possible to represent the one in conjunction with the other is what renders the world (and everything it contains) *certain*.

The Unified Model of the Universe

The model of the Universe, the matrix of worlds, is the absolute, complete, and certain set of all aspects of all worlds.

Like the model of the world, the model of the Universe cannot be conceived without taking into account the conscious representation of states and the a-conscious representation of events. It is the conjunction of these two.

The Universe and consequently its model would only be one of its own aspects were it not the All from which we distinguish each event and each possible world state. It is necessarily outside time (and outside gravity) and is the structured, unified set of states and the indistinct, unified set of events, whilst also being the fundamental origin of each.

The conscious representation of the Universe is thus the complete set of states from which it is possible to distinguish any (causal) relationship, and its a-conscious representation is the complete set of events. And the first event to be distinguished would necessarily be a big bang.

The joint representation of the conscious structure and the a-conscious unity of the Universe constitutes its model.

It can only be the model the Universe on the joint condition that its structure is not composed of elements found in each manifested state or, in other words, constituting their absolute reality, and that it be certain that the indefinite form of the complete set of events be perfectly circumscribed by the structure without omitting any event whatsoever.

Hence everything which is in the Universe is in the model, and everything that the model reveals is in the Universe.

The mental representation of the Universe so conceptualised with a geometry guaranteeing a perfectly regular structure is thus certain, complete, and absolute. The model is *perfect*.

It is the Universe.

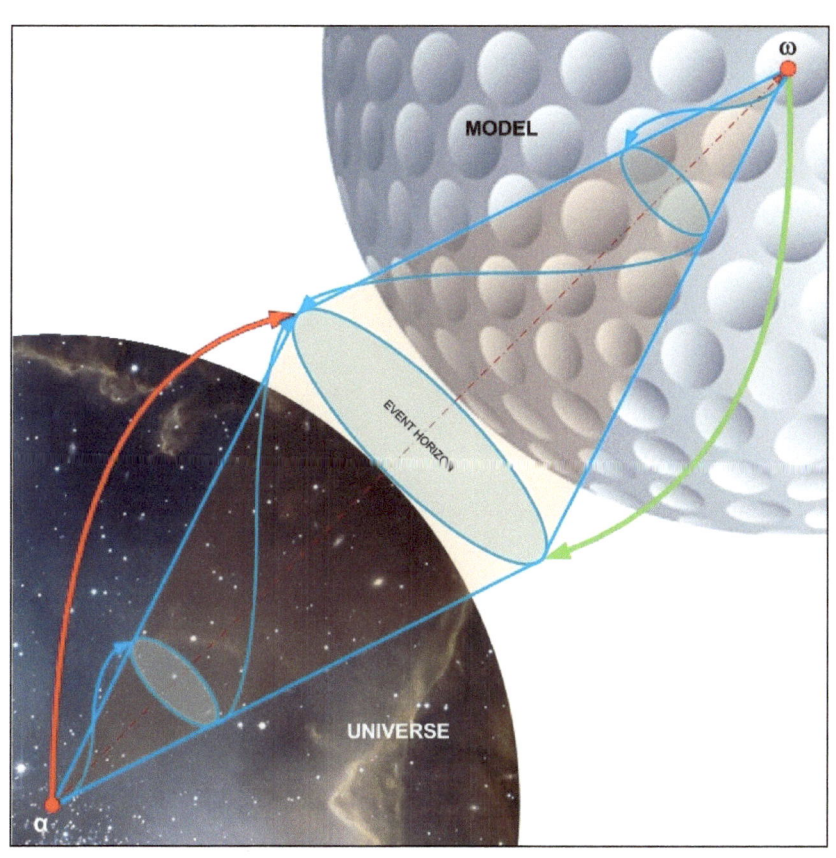

STATEMENT OF PRINCIPLES

• We can imagine many worlds, but there is only one Universe in which we can exist and which our reasoning can access. Our world is a product of this.

• The Universe is made up of 5 dimensions and five only, two of which are employed to show us the world: time and gravity. These are not dimensions which are native to the Universe.

• Consequently, none of the Fundamental Laws can include them in their equations.

• Only the Universe has certain Laws. They are omnipresent and unique.

• They make themselves known through both their presence and their absence. Everything is 'designed': the Universe knows neither teleonomy, nor evolution.

• We form two representations of the Universe and the world: a conscious one, consisting of our perceptions and concepts which we infer from them, the other is a-conscious (and therefore 'unreal'), fed by the same perceptions and concepts that form the first.

• The architecture of the a-conscious representation forms progressively during neuronal ontogeny and stabilises in the 'Age of Reason' in a primitive, stable – and therefore universal – form, supporting the Intellect Agent.

• The combination of the two allows us to accept a common model of the Universe, certain mental representation, complete and absolute. Everything that is in the Universe is also in the model, and everything that the model reveals is in the Universe. It is the Universe.

• All possible discoveries are visible in the model without the need to carry out experiments. The perfection of the model, the true source of scientific inspiration, proves them.

• The Universe and its mental model are two developments from the same prime, guiding pattern, constructed in a primitive form, the matrix of worlds.

• Everything, from the infinitely small to the infinitely large – where the passage from one order to another higher order can be marked by ruptures – stands in a unique, fractal sequence which paves the way for a **Unified Theory**.

79

www.ingramcontent.com/pod-product-compliance
Lightning Source LLC
Chambersburg PA
CBHW041102180526
45172CB00001B/75